LANDSLIDES

By Pamela McDowell

LIGHTBOX
openlightbox.com

LIGHTBOX

Go to
www.openlightbox.com,
and enter this book's
unique code.

ACCESS CODE

LBU29572

Lightbox is an all-inclusive digital solution for the teaching and learning of curriculum topics in an original, groundbreaking way. Lightbox is based on National Curriculum Standards.

STANDARD FEATURES OF LIGHTBOX

 AUDIO High-quality narration using text-to-speech system

ACTIVITIES Printable PDFs that can be emailed and graded

 SLIDESHOWS Pictorial overviews of key concepts

VIDEOS Embedded high-definition video clips

WEBLINKS Curated links to external, child-safe resources

 TRANSPARENCIES Step-by-step layering of maps, diagrams, charts, and timelines

 INTERACTIVE MAPS Interactive maps and aerial satellite imagery

QUIZZES Ten multiple choice questions that are automatically graded and emailed for teacher assessment

 KEY WORDS Matching key concepts to their definitions

Copyright © 2022 Smartbook Media Inc. All rights reserved.

Contents

Lightbox Access Code 2

Deadly Landslides Occur with Little Warning 4

Most Landslides Are Caused by Heavy Rainfall 6

Some Landslides Can Be Prevented 8

Landslides Are Common Across the World 10

All-Time Records 12

Landslides in the United States 14

Most of the Philippines Is at Risk for Landslides 16

Scientists Use Many Tools to Study Landslides 18

Little-Known Facts 20

Landslides Are Most Dangerous at the
Leading Edge 22

Landslides Can Be Part of a Series of Disasters 24

Ancient Peoples Believed Gods Shook the Land,
Creating Landslides 26

Landslide Timeline 28

Test Your Knowledge 29

Create a Disaster Kit 30

Key Words/Index 31

Log on to www.openlightbox.com 32

Deadly Landslides Occur with Little Warning

A landslide is the movement of dirt, rocks, and **debris** down a slope. Landslides are most common in coastal or mountain regions. They occur many times each year and can happen in almost all parts of the world. In general, steeper slopes have a greater risk of landslides than more gradual ones.

Landslides are more likely when natural forces or human activities make an area of land less **stable**. Dirt and rocks tend to slide downward on a slope that is not stable. A large landslide can change the shape of a mountainside. Sometimes, the unstable land will slide away very quickly. Fast-moving rocks and mud can destroy entire towns, killing people in the slide's path. Often, the deadliest landslides are the ones that happen with little warning. Scientists around the world study landslides to learn how to predict and prevent them.

The cleanup after a landslide often requires heavy equipment.

People were told to leave a YMCA center in Japan before heavy rains caused landslides in 2014. Most of the center's parking area was destroyed.

Most Landslides Are Caused by Heavy Rainfall

Heavy rainfall is the main cause of most landslides. Other natural processes, such as **erosion**, can help make land unstable. Large landslides often follow other natural disasters, such as earthquakes. Floods, fires, and volcanic eruptions can also set off landslides. Sometimes, people help to start them by changing the landscape.

CAUSES OF LANDSLIDES

RAIN AND EROSION	EARTHQUAKES	OTHER NATURAL DISASTERS	HUMAN ACTIVITY
• During heavy rainfall, land takes in a great deal of water. • The soil becomes heavier. It may then slide down hillsides or shorelines. • Erosion helps to make land more likely to slide downward.	• Earth's surface is divided into large sections called plates. They are always moving. • The plates push or scape one another, causing earthquakes. • The shaking can make land unstable, causing a landslide.	• Floods and fires remove the plants from an area. • Without plant roots to keep it in place, the land on hillsides may be unstable. • Volcanoes also cause landslides. Hot gases can weaken rocky hillsides.	• Mining can make land less stable. • Digging into hillsides and using explosives can loosen dirt and rock. • Clearing land for homes or farms can also increase the risk of landslides, by removing trees.

Shear strength is a measure of a slope's ability to stay in place. It is the result of two forces, called cohesive and frictional strength. **Particles** of different materials, such as rock, clay, and sand, bond together in the ground. This bonding provides cohesive strength. The mixing of different materials also makes an area of land able to resist movement. This resistance is frictional strength. Together, cohesive and frictional strength can be strong enough to resist **gravity**.

Natural or human actions can change the mix of materials in soil. They can also weaken the bond. The weak surface can slide downward. Sometimes, the surface slides slowly. This is called a creep. Special equipment measures creeps.

A landslide can block roads, leaving people stranded.

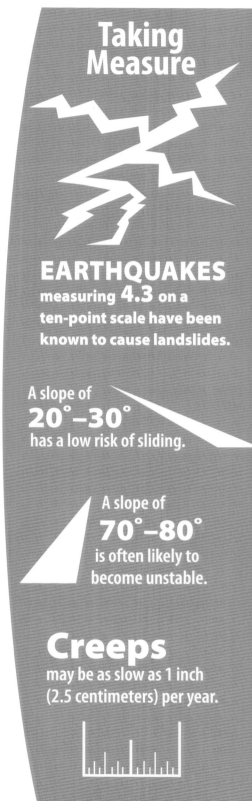

Taking Measure

EARTHQUAKES measuring **4.3** on a ten-point scale have been known to cause landslides.

A slope of **20°–30°** has a low risk of sliding.

A slope of **70°–80°** is often likely to become unstable.

Creeps

may be as slow as 1 inch (2.5 centimeters) per year.

LANDSLIDES 7

Some Landslides Can Be Prevented

Scientists have ways to strengthen slopes and help prevent landslides. In some places, they can use machines to reshape land, making it less steep. Sometimes, they can build special walls called retaining walls. These walls catch loose debris and keep it in place, or retain it. Plants, chemicals, and drains can help save slopes in danger. Certain types of plants, when added to a slope, can hold down the top layer of the land even when a slide starts to occur. There are chemical sprays that can make a slope stronger. Drainpipes can direct water away from a slope that is ready to slide.

It is not possible to stop every mountainside from crashing down. Yet, some destruction can be prevented, and lives can be saved. Where landslides are likely, there are ways to lessen the effects. Workers can direct debris away from settlements by building pathways for it. People can be kept out of dangerous areas. To guard against destruction, some areas have restrictions. For example, laws prevent neighborhoods from being built on or near unstable land.

After landslides in Japan in 2014, the government began to survey areas at risk and send out warnings as needed.

Many landslides can be predicted. Scientists can monitor, or watch for, heavy rainfall. They can measure slope strength. In these ways, they can try to predict where a hillside may slide and alert people nearby. This science was used in Utah in 2013. Scientists monitored ground movement in Bingham Canyon. They studied the land around an open-pit copper mine. They knew a slide would happen.

When a landslide warning went out, workers moved power lines and a cell tower. They evacuated, or left, the area where the slide was expected. On April 10, 2013, the land came crashing down. About 147 million tons (133 million metric tons) of land fell downward. It was the largest known slide in North American history not caused by a volcano. The huge landslide destroyed some machinery, but no people died or were injured.

In areas at risk for landslides, officials post warnings.

The Bingham Canyon landslide caused 16 small earthquakes.

LANDSLIDES 9

Landslides Are Common Across the World

Landslides have happened since prehistoric times. The effects of past landslides can still be seen. More than 10,000 years ago, debris from the Saidmarreh landslide in southwestern Iran created a dam that blocked a river. A lake formed behind this dam. Layers of **sediment** gathered in the lake until the dam broke. After the lake's water drained, the sediment left behind created fertile farmland. The land there remains productive.

Areas in southern and eastern Asia that receive heavy rainfall from **monsoons** and **typhoons** are at special risk. India's monsoon rains occur from June to October. In June 2013, floods caused by heavy rain produced landslides in Kedarnath, India. It is estimated that more than 5,700 people died.

A week after landslides near Rio de Janeiro, Brazil, some victims still could not be reached.

Other regions with rainy seasons experience more landslides than areas where rainfall occurs more evenly during the year. The area of Brazil around the city of Rio de Janeiro has a rainy season from January to April. In early January 2011, the area received 10 inches (25 cm) of rain in just 24 hours. A series of mudslides destroyed many communities, and more than 900 people lost their lives. Some of the largest landslides have hit places that were soaked by water and shaken by earthquakes or volcanoes. When a volcano erupted in Colombia in 1985, it melted ice in the mountaintops. More than 23,000 people died in mudslides caused by the sudden melting.

Some scientists estimate that, on average, more than 8,000 people die in landslides around the world each year. This is an average, so the numbers are higher in some years. In 1999, strong rains hit the coastal area of Vargas, Venezuela. Landslides destroyed communities along the coastline. From 10,000 to 30,000 people were buried under mud or swept out to sea.

In Kedarnath, India, officials made posters showing missing people to try to find them.

25–50
Number of people killed in landslides each year in the United States.

$3.5 billion
Approximate cost of U.S. landslide damage each year.

57 Number of people who died in the 1980 eruption of the Mount St. Helens volcano in Washington, many of them as a result of a huge landslide caused by the eruption.

MORE THAN 30,000
Number of people who died worldwide in landslides in 2005, the highest number for any year in the 21st century.

All-Time Records

Large storms often cause major landslides. Slides that occur in populated areas tend to cause the most expensive damage. In these places, there are many buildings and people in a small area of land. Sometimes, the number of people injured or killed is high.

MOST EXPENSIVE
Damage estimates after the 2013 landslide in Bingham Canyon, Utah, ranged from $770 million to $1 billion, making it the costliest in U.S. history.

DEADLIEST
In Oso, Washington, 43 people died in a landslide in 2014. This was the highest U.S. death toll from a landslide not related to another natural disaster, such as a volcanic eruption. Logging had removed many trees from the slope that slid.

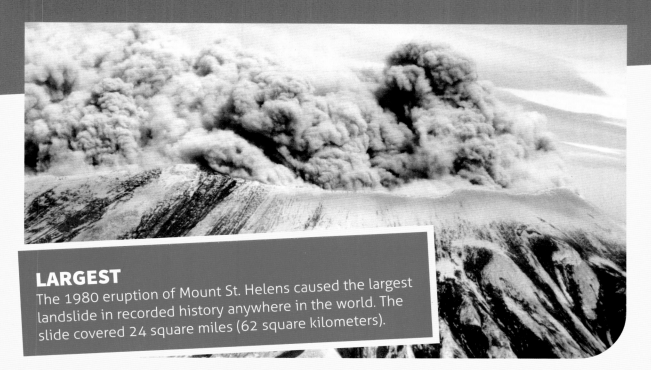

LARGEST
The 1980 eruption of Mount St. Helens caused the largest landslide in recorded history anywhere in the world. The slide covered 24 square miles (62 square kilometers).

MOST UNUSUAL
Scientists believe that a landslide on Mars was caused billions of years ago when the planet was hit by a giant **asteroid**. The area on the surface of Mars that slid was about the size of the United States.

LANDSLIDES 13

Landslides in the United States

Landslides occur in all parts of the United States. In the Appalachian Mountains near the East Coast, landslides are common. They also often occur in the coastal mountain ranges of the Pacific Northwest. The U.S. government creates maps showing areas at high risk in the 48 **contiguous** states.

AREAS AT RISK FOR LANDSLIDES
- higher risk
- lower risk

MAP SCALE
0 — 500 miles / 500 kilometers

1 **Location:** Oso, Washington
Date: March 22, 2014
Deaths: 43

2 **Location:** La Conchita, California
Date: January 10, 2005
Deaths: 10

14 FORCES OF NATURE

3 **Location:** Gros Ventre Wilderness, Wyoming
Date: June 23, 1925
Deaths: 6

4 **Location:** Thistle, Utah
Date: April 1983
Damage: The entire town was destroyed, though people had time to leave safely.

Most of the Philippines Is at Risk for Landslides

Countries in East and Southeast Asia where landslides are common include China, Indonesia, and the Philippines. Landslides in the Philippines have been especially deadly. Much of the land in the Philippines is mountainous or hilly. The island nation often has earthquakes and is hit by typhoons. Up to 80 percent of the Philippines is at risk of landslides.

Rescue workers in the Philippines use devices called life detectors to find survivors who may be buried under landslides.

The population of the Philippines is more than 107 million people. The population is growing, and cities have expanded rapidly. Many millions of people now live in areas where landslides are likely. In the capital city of Manila and communities around it, almost 3 million people live in unstable areas.

The worst landslide in the Philippines in recent history occurred in February 2006. Two weeks of heavy rain fell on the Visayan Islands in the central part of the country. The rain combined with a minor earthquake to start a landslide that buried an entire village. More than 1,000 people died.

Strong typhoons have hit the Philippines in recent years. Typhoon Parma caused landslides in the northern Philippines in 2009, killing more than 160 people. The heavy rains from typhoon Haiyan in 2013 also produced landslides.

After the 2006 landslides in the central Philippines, photos taken from helicopters showed the extent of the disaster.

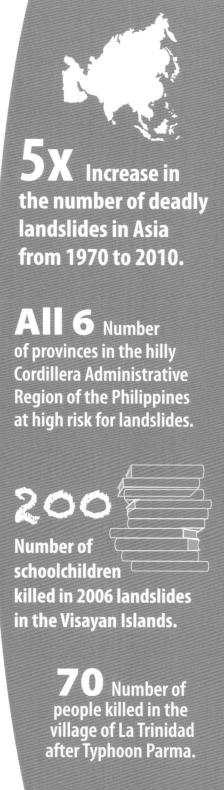

5x Increase in the number of deadly landslides in Asia from 1970 to 2010.

All 6 Number of provinces in the hilly Cordillera Administrative Region of the Philippines at high risk for landslides.

200 Number of schoolchildren killed in 2006 landslides in the Visayan Islands.

70 Number of people killed in the village of La Trinidad after Typhoon Parma.

LANDSLIDES 17

Scientists Use Many Tools to Study Landslides

Scientists who study Earth's land and history are called geologists. The U.S. Geological Survey (USGS) studies areas where landslides are likely to occur. Its geologists set up equipment to measure rainfall and soil moisture in at-risk locations. Devices called sensors detect even tiny underground movements at these remote, or distant, monitoring stations.

Remote monitoring stations transmit, or send, information to the USGS using wireless communication systems. The USGS receives additional data, or information, from **satellites** in space that track weather. This data can help geologists predict when an area at risk for landslides is likely to receive heavy rain.

In an area at risk for mudslides, officials can position concrete K-rails to direct mud away from homes.

The USGS monitors some sites continuously. The stations in some places send data every five minutes. The slope and soil are never exactly the same at any two locations. The amount and type of vegetation, or plant life, can vary greatly. Scientists consider all of these factors to predict how much moisture an area can absorb, or take in. They study which conditions caused landslides in the past. The amount of water that an area can hold is called a rainfall threshold. Rainfall above this threshold may start a landslide.

In California, the USGS places sensors on slopes where fires have burned the vegetation. At first, the sensors monitor the slope continuously. As new plants grow, the hillside becomes more stable. Then, the data are gathered less often. Eventually, geologists move the sensors to another location where monitoring is needed.

AREAS MONITORED FOR LANDSLIDE RISK IN SOUTHERN CALIFORNIA

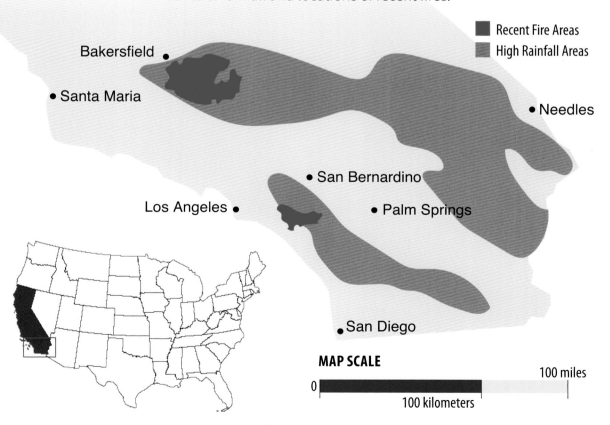

In 2012, the USGS chose certain areas to monitor based on the amounts of rainfall and locations of recent fires.

Little-Known Facts

SPIDEY SENSE

The USGS uses special devices called spider units to study landslides that have already begun. Helicopters drop spider units on landslides in motion. Spider units provide data that can warn workers of changes in the slide's force or direction.

IN NEAR TIME

Most scientists call the use of up-to-the-minute data "real-time monitoring." Others say it is more like "near time." Information transmitted from remote monitoring stations may be sent to satellites in space, which then relay the data to USGS scientists. The time it takes for data to travel is brief. However, there is some delay.

DOUBLE DANGER

A pyroclastic flow occurs during a volcano's eruption. Made up of ash, **lava**, rocks, and gas, it can be as hot as 1,300° Fahrenheit (700° Celsius). It sets landslides in motion. It also melts any snow or ice that was at the top of the volcano. When landslides pick up water, they become lahars, which are made of water, mud, and rock. These deadly flows look like rivers of wet cement.

WARNING SIGNS

In the moments before a landslide, there may be new cracks in pavement or swells on the ground. Leaning telephone poles and fences may mean land has started to move. Windows and doors may stick because their frames are no longer straight.

LEARNING FROM LANDSLIDES

Waste from a coal mine was piled above the village of Aberfan, Wales. On October 21, 1966, heavy rains caused the pile to slide, burying a school filled with children. After the deadly landslide, new laws were passed, forcing mining companies to monitor their waste.

Landslides Are Most Dangerous at the Leading Edge

As a landslide travels down a mountain or hillside, it picks up materials in its path. The items can include boulders, trees, and even homes. The objects break up and become debris. Much of this debris stays at the front of the landslide, or the leading edge. It becomes thicker and heavier as it travels downhill. In a landslide caused by heavy rain, wet mud may build up behind the leading edge. Eventually, the mud may break through. The mud then travels down the hillside more quickly.

Workers searched a river for victims after a 2014 landslide near Hiroshima, Japan.

As a landslide moves, it picks up speed. Some landslides have been reported moving at speeds of almost 200 miles (320 km) per hour. This rapid movement makes a wind strong enough to tear leaves from trees. It creates **momentum** that can drive the slide far into a valley at the foot of the slope. Sometimes, a slide's momentum is even strong enough to drive the leading edge up a mountain slope on the opposite side of the valley.

After a landslide, cleaning up mud and debris can be an expensive task for local governments.

In the United States, sliding rock and debris cause the most deaths and injuries from landslides. Broken electrical, sewage, and gas lines cause additional losses. If it is not possible for people to escape the leading edge of a landslide, the safest place they can go is often the top floor or roof of a house or other building.

When a landslide includes a great deal of mud, it may be called a mudslide.

Landslides Can Be Part of a Series of Disasters

Landslides that follow other disasters, such as volcanic eruptions, can be more destructive than the events that triggered, or started, them. Sometimes, a landslide sets off another natural disaster. In 1963, there was a landslide at Monte Toc, Italy. Millions of cubic yards (cubic meters) of rock fell down the mountainside. The rock fell into a human-made lake with such force that it created a huge wave. The 820-foot (250-m) tall wave caused a flood. More than 1,900 people died.

When the top of Monte Toc broke off in 1963, it crashed into a reservoir of water created by a dam.

EFFECTS OF LANDSLIDES

Tsunamis

Sometimes, landslides happen under the ocean following an earthquake. These landslides can cause huge waves of seawater called tsunamis. In 1998, an earthquake caused an underwater landslide near the coast of New Guinea, a large island north of Australia. The landslide triggered a tsunami that wiped out several villages in the country of Papua New Guinea. Thousands of people lost their lives.

Flooding

Rock and debris from a landslide can block rivers. These dams can create new lakes, but they are unstable. Dams made up of debris are likely to shift and break, causing flooding of the areas downhill. In 1941, a landslide dammed the Santa River in Huaraz, Peru. When the dam failed, the resulting flood swept everything downhill toward the sea. As many as 6,000 people died.

Ongoing Damage

A landslide tears up or covers everything in its path. It can kill animals and cover or pull up all vegetation in an area. Lakes and parts of rivers may be covered with rock and debris. Sometimes, entire **ecosystems** are destroyed. It can take years for new vegetation to grow and animal life to return to an area severely damaged by a landslide.

Ancient Peoples Believed Gods Shook the Land, Creating Landslides

Before modern scientists studied landslides, people in ancient cultures created myths, or stories, to explain why these disasters happened. In ancient Greek mythology, the god of the sea was Poseidon. He carried a trident, a spear that looks like a huge fork. The Greeks called Poseidon the "Earth Shaker." They believed that by banging his trident on the surface of Earth, Poseidon caused landslides, earthquakes, and floods.

In Malawi, Africa, ancient stories told of a large underground snake named Napolo. He lived in the mountains but sometimes moved to lower land. When Napolo moved, huge blocks of dirt and rock would fall from the mountains.

An American Indian legend told of a spirit, or ghost-like being, in the form of a giant toad or frog. It lived below Lituya Bay in southeastern Alaska. According to the traditional beliefs of the Tlingit people, if the peace of the bay was disturbed, the spirit shook the land and sea.

Poseidon was called Neptune by the ancient Romans. He was often shown in Roman artworks.

THE BRIDGE OF THE GODS

A traditional story of American Indian groups in the Pacific Northwest describes the creation of rapids in the Columbia River. Rapids are areas where the water flows quickly over rocks in the river. The god Tyhee Sahale and his two sons traveled down the Columbia, or Great River, to settle in a fertile land. The sons fought over the land. Sahale shot an arrow far to the north and another far to the west. He said to his sons, "Go. Find the arrows. Where they lie, you shall have the land." Sahale then raised mountains between his sons, so their tribes would not fight. The Great River between them became a sign of peace. Sahale built a stone bridge over the river, so the tribes could be friends.

Both sons fell in love with the same goddess. They fought over her, and their people fought. This angered Sahale. He shook the land, and great boulders slid from the mountains, covering villages. The stone bridge collapsed into the Columbia River, creating the rapids seen there today.

Landslide Timeline

1200

1248
Five villages are destroyed when a landslide of limestone rock falls from Mont Granier, France.

1800

1806
Landslides fall into Lake Lauerz, Switzerland, causing a large wave that destroys four villages.

1868
The largest earthquake to strike the Hawai'an islands causes landslides, a tsunami, and a volcanic eruption.

1903
In 100 seconds, 90 million tons (82 million metric tons) of rock crash down Turtle Mountain onto the town of Frank, Alberta. Seventy people die in the largest and deadliest landslide in Canadian history.

1900

1919
More than 100 villages are destroyed by lahars caused by the eruption of Mount Kelud in India.

2014
A day of heavy rain in the region around Hiroshima, Japan, is followed by landslides the next morning. Although there were warnings to leave at-risk areas, 72 people die.

2000

2007
In Chittagong, Bangladesh, rain falls on a hillside being cut to create level land for farming. A landslide results, killing 127 people.

28 FORCES OF NATURE

Test Your Knowledge

1. What kind of material caused a landslide in Aberfan, Wales, on October 21, 1966?

A. Waste from a coal mine

2. What two forces combine to give a slope shear strength?

A. Cohesive strength and frictional strength

3. What is the main cause of landslides?

A. Heavy rainfall

4. Where did the Saidmarreh landslide occur more than 10,000 years ago?

A. Southwestern Iran

5. How many people died worldwide in landslides in 2005?

A. More than 30,000

6. What do scientists think caused a landslide on Mars billions of years ago?

A. An asteroid

7. How much of the Philippines is at risk of landslides?

A. Up to 80 percent

8. What is the name for the amount of water that land can hold before there is a risk of landslides?

A. Rainfall threshold

9. What type of scientist studies Earth's land and history?

A. A geologist

10. In ancient Greek mythology, which god caused landslides?

A. Poseidon

LANDSLIDES 29

Create a Disaster Kit

Disasters happen anytime and anywhere. When an emergency happens, you may not have much time to respond. The Red Cross says that one way to prepare is by assembling an emergency kit. If you and your family have gathered supplies in advance, you will be prepared in case it is necessary to evacuate or stay at home for a time until help arrives.

1 Know the risk of landslides in your area. Watch for changes in the flow of water and the shape of the land.

2 Make an evacuation plan. What is the quickest route to safety? How can you get there? Have a back-up plan in case roads are blocked.

3 Know which radio stations, television stations, or websites will provide up-to-date information about disasters.

4 Put together an evacuation kit with at least three days worth of supplies that you will take with you.

5 Gather supplies for a home kit. These supplies should be able to support you and your family for at least two weeks if you cannot leave the area.

6 Always have a detailed emergency plan and practice it with your family. Everyone should know where to go and how to contact each other in an emergency.

What You Need
- water
- food
- flashlight
- radio
- extra batteries
- first-aid kit
- phone numbers and addresses of family members and friends
- soap and toilet paper
- cell phones and chargers
- cash
- blankets
- warm clothing
- maps of your area

Key Words

asteroid: a rocky object that orbits, or travels around, the Sun

contiguous: connected to or touching one another

debris: pieces of something that has been destroyed

ecosystems: communities of living things and resources

erosion: wearing away of something by water or wind

gravity: a force that pulls objects toward the center of Earth

lava: very hot melted rock that flows out of a volcano

momentum: the force of something in motion that tends to make it keep moving

monsoons: steady winds that bring moist air to a region at a certain time of year, causing long periods of heavy rain

particles: small pieces of a substance

satellites: human-made objects that are sent into space and orbit Earth

sediment: rock, sand, and dirt that has settled at the bottom of a body of water

stable: firm and not easily moved from its location

typhoons: storms with high wind and heavy rain that occur in the western Pacific Ocean

Index

Aberfan, Wales 21, 29

Bingham Canyon 9, 12
Brazil 10, 11

creep 7

earthquake 6, 7, 9, 11, 16, 17, 25, 26, 28
erosion 6

Frank, Alberta 28

India 10, 11, 28
Italy 24

Japan 5, 8, 22, 28

lahar 21, 28

Mars 13, 29
mining 6, 9, 21, 29
monsoon 10
Mount St. Helens 11, 13

Oso, Washington 12, 14

Philippines 16, 17, 29
Poseidon 26, 29

rainfall threshold 19, 29

Saidmarreh 10, 29
shear strength 7, 29

tsunami 25, 28
typhoon 10, 16, 17

U.S. Geological Survey (USGS) 18, 19, 20

volcano 6, 9, 11, 12, 21, 24, 28

LIGHTBOX

➕ SUPPLEMENTARY RESOURCES

Click on the plus icon ➕ found in the bottom left corner of each spread to open additional teacher resources.

- Download and print the book's quizzes and activities
- Access curriculum correlations
- Explore additional web applications that enhance the Lightbox experience

LIGHTBOX DIGITAL TITLES
Packed full of integrated media

VIDEOS

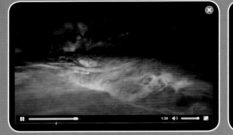

INTERACTIVE MAPS

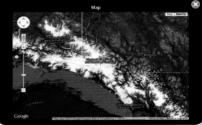

WEBLINKS

SLIDESHOWS

QUIZZES

OPTIMIZED FOR
- ✓ TABLETS
- ✓ WHITEBOARDS
- ✓ COMPUTERS
- ✓ AND MUCH MORE!

Published by Smartbook Media, Inc.
276 5th Avenue Suite 704 #917
New York, NY 10122
Website: www.openlightbox.com

Copyright © 2016 Smartbook Media, Inc.
All rights reserved. No part of this publication may be reproduced, stored in a retrieval system, or transmitted in any form or by any means, electronic, mechanical, photocopying, recording, or otherwise, without the prior written permission of the publisher.

Library of Congress Cataloging-in-Publication Data
McDowell, Pamela.
 Landslides / Pamela McDowell.
   pages cm -- (Forces of nature)
 Includes index.
 ISBN 978-1-5105-0082-2 (hard cover : alk. paper)
 -- ISBN 978-1-5105-0306-9 (soft cover : alk. paper)
 -- ISBN 978-1-5105-0083-9 (multi-user ebook)
 1. Landslides--Juvenile literature. 2. Natural disasters--Juvenile literature. I. Title.
  QE599.A2M34 2015
  551.3'07--dc23
                     2014038991

Printed in Guangzhou, China
3 4 5 6 7 8 9 0 26 25 24 23 22

042022
120422

Project Coordinator: Aaron Carr
Art Director: Terry Paulhus

Every reasonable effort has been made to trace ownership and to obtain permission to reprint copyright material. The publishers would be pleased to have any errors or omissions brought to their attention so that they may be corrected in subsequent printings.

The publisher acknowledges Getty Images as its primary image supplier for this title.